Management Division 2017

Core Programming Area at the 2017 AIChE Spring Meeting and
13th Global Congress on Process Safety

San Antonio, Texas, USA
26 – 30 March 2017

ISBN: 978-1-5108-4143-7

Printed from e-media with permission by:

Curran Associates, Inc.
57 Morehouse Lane
Red Hook, NY 12571

Printed by Curran Associates, Inc. (2017)

For permission requests, please contact AIChE
at the address below.

AIChE
120 Wall Street, FL 23
New York, NY 10005-4020

Phone: (800) 242-4363
Fax: (203) 775-5177

www.aiche.org

Additional copies of this publication are available from:

Curran Associates, Inc.
57 Morehouse Lane
Red Hook, NY 12571 USA
Phone: 845-758-0400
Fax: 845-758-2633
Email: curran@proceedings.com
Web: www.proceedings.com

TABLE OF CONTENTS

482030 AGILE Award Keynote Address

Abstract:

Factors in Selecting AGILE Honoree

1. Overall sustained excellent industry leadership in businesses that employ chemical engineers

2. Recognized for efforts in innovation and creating strategies that ensure success

3. Value of the keynote to the meeting attendees and the members of the Institute Schedule

Event	Time	Location
Continental Breakfast	7:00AM - 8:00AM	Henry Gonzalez Convention Center, Lila Cockrell Theatre
AGILE Keynote Address	8:00AM - 9:15AM	Henry Gonzalez Convention Center, Lila Cockrell Theatre

The 2017 AIChE Spring Meeting and 13th GCPS AGILE Keynote Address will be presented by AGILE award recipient John Y. Televantos, Partner, Arsenal Capital Partners.

John Televantos is a Partner at Arsenal Capital Partners, a middle market private equity firm focused on the specialty industrial, healthcare and specialty financial service sectors. Before joining Arsenal in 2006, Dr. Televantos was President of Aqualon Company, owned by Hercules, Inc.; President and Founder of Helios Chemical Company; and Chief Executive Officer for Foamex International.

Dr. Televantos began his career with Union Carbide in 1977 in silicones research. There he developed two key processes for the manufacture of Methyl Chloride and Siloxane polymers that were commercialized. These processes not only improved efficiency and reduced cost but also dramatically reduced waste. He moved through a variety of management positions there and later at ARCO Chemical, where he became vice president of strategy and development.

At ARCO John co-invented and led a team that developed and commercialized a breakthrough technology in polyol manufacture (a key raw material for polyurethanes) eliminating a whole process step along with waste and emissions, while at the same time opening the door to the creation of new higher-performance polyurethanes. The technology has since been licensed by most of the major global polyol manufacturers with significant waste reduction worldwide.

At Arsenal Capital, Dr. Televantos serves as co-head of the specialty industrial practice. During his tenure, he has been part of several acquisitions and successful exits in the chemicals and materials sectors. In addition to serving on the boards of several Arsenal-owned companies, he is lead director on the board of Balchem Corp.

481492 How Could Being a PE Help Me?

Abstract:

This presentation discusses the advantages of professional engineering licensure, how engineering licensure practices are based on state code and how they vary from state to state, and NCEES initiatives to enhance licensure mobility

481908 Benefits for Having a PE According to a Young Professional

Abstract:

Earning a PE can do a wide variety of things for a person's career. It can help to show that they are knowledgeable about chemical engineering and to open up many future possibilities. These benefits apply to both younger and older engineers, but are especially critical to younger professionals.. In this discussion, I will walk through why as a young professional hoping to ascend the career management ladder, I became and will maintain my PE license.

481912 How a PE Enhances Career Success Potential

Abstract:

Jim Porter is an example of an engineer who had a successful career and was not a licensed PE. However, he feels that being licensed is important in today's world, and he will outline his reasoning in his remarks

482664 New Manager Prospectives from Johnny Rokisky: Fungicide Formulation Production Superintendent from Syngenta

Abstract:

Johnny Rokisky is the Unit Superintendent of a Fungicide Production Unit for Syngenta in Houston Texas. He received his bachelors of Chemical Engineering from Virginia Tech in 2010 and joined Syngenta after in a Leadership Development Rotational Program. During this 2 years in the rotational program, he worked as a production engineer, a formulation engineer, and a Supply Chain Planner. After graduating from the rotational program, he moved to Omaha to work as a new product engineer. Shoftly after, he was promoted to become a Unit Superintendent at the Houston site.

482665 New Manager Prospectives from Carol Schmidt: A Technical Sales Manager for the Sgl Group

Abstract:

Carol works for SGL Group, as a Technical Sales Manager for the Process Technology/Engineered Solutions department. SGL's product lines focus on non-metal equipment for corrosive services. Mainly, this is heat exchangers with graphite or silicon carbide internals, PTFE bellows and lined piping/columns, and acid synthesis systems. Carol has been doing this for 1.5 years. Carol does not have any direct employee reports, but she manage 12 rep agents in 5 states. Carol is responsible for a sales territory spanning 12 states and she travel about 2-3 weeks of the month. Her responsibilities include, but are not limited to, regularly visiting customers, finding opportunities, forecasting, budgeting, quotations, sales orders, and negotiation.

Carol will speak about managering her group indirectly, prioritizing her workload, and handling all of this dealing with gender/age/ethnic issues as well as children.

482666 New Manager Prospective Carlos Andres Salamanca

Abstract:

Carlos is the current STS-AIChE Chair Elect for 2017, former STS-AIChE Treasurer for 2015 and 2016. He has been an active member of the STS AIChE Young Professionals Board serving as Secretary, Treasurer, Vice-Chair and Chair. He also has represented AIChE with the Offshore Technology Conference events including OTC Houston, Arctic Technology Conference, OTC Brasil, and OTC Asia in different roles. Carlos is also active with Toastmasters, SPE (Society of Petroleum Engineers), and the PMI (Project Management Institute) Houston Chapter.

He has held several positions through different oil and gas projects. Carlos has been working with oil companies in Latin America with focus on Brazil, Colombia and Mexico. His most recent role was with National Oilwell Varco working with challenging projects including oil and gas drilling and production rigs. He also enjoys teaching different subjects. Carlos started his career with Texaco, now Chevron, at the lubricants plant in Bogotá, Colombia.

Carlos earned a BS in Chemical Engineering from the Universidad Nacional de Colombia, and received an MBA from the Bauer College of Business at the University of Houston. Carlos holds a PMP (Project Management Professional) certificate from the Project Management Institute, and an EIT (Engineer-In-Training) certificate from the Texas Board of Professional Engineers. His professional interests focus on Oil and Gas Economics, Project Management, Global Business, Latin American Studies and new Engineering technologies. Carlos likes to read major business publications, leadership, self-improvement, public speaking and management books. He enjoys spending time with family and friends, all sporting events, riding his bicycle, and traveling. Carlos is also fluent in Spanish and is currently learning Portuguese.

481240 Critical Leadership Moments

Abstract:

When situations at work turn from casual to critical, we instinctively react poorly. Our logic center shuts down and we resort to primitive thinking skills. A critical moment is defined by three components which set it apart: 1) high stakes are involved, 2) there are opposing opinions, and 3) strong emotion is present. If not all three of these elements are recognized, employees always silently "act out" their frustrations in negative ways. Sometimes unknowingly. As managers, when you are taught how to "talk through" high-stake, emotionally charged topics, the outcome is dramatically more successful on the performance of individuals, teams, and entire organizations.

480857 Your Career Path Is like a Distillation Column

Abstract:

A typical distillation column will have its distillate stream at the top and residue stream at the bottom. No matter which stage your feed stream is coming from and what the composition og your feed stream is, all the molecules should select a right steam to go. Your career path is same. Normally there are two kinds of career paths(managerial career path and technical/professional career path) that you can follow in your career development. No matter what your level of KSA(knowledge, skill and ability) is when you graduate from your college, you should pick up a right career path to follow. Different career path will require different KSA development so it is better for you to make a decision as early as possible. This presentation will provide you're the detailed comparison of two career paths and your development approach of required KSA.

AUTHOR INDEX